Science Inquiry

Weather

by Joe Baron

Science Inquiry

Science in a Snap!

Explore Activity

▶ **Question:** What can you observe about the temperature of water placed in sunlight?

Directed Inquiry

▶ **Question:** What happens over time to water in an open cup and in a covered cup?

Directed Inquiry

▶ **Question:** How does the weather change from day to day?

Science in a Snap!

Next Generation Sunshine State Standards
SC.2.E.7.2 Investigate by observing and measuring, that the Sun's energy directly and indirectly warms the water, land, and air.

Put a dime in sunlight. Put another dime in the shade. Wait about 15 minutes. Touch the dime in the shade. Then touch the dime in sunlight. What is different about the dimes? What do you think caused the difference?

4

Season to Season

Next Generation Sunshine State Standards
SC.2.E.7.1 Compare and describe changing patterns in nature that repeat themselves, such as weather conditions including temperature and precipitation, day to day and season to season.

Make a book about seasons. Use 4 sheets of paper. On each paper, write the name of a different season. Draw a picture of the weather during that season. Write words to tell about the seasons. Talk with a partner and compare your drawings.

How Hot Is It? How Cold Is It?

Next Generation Sunshine State Standards
SC.2.E.7.1 Compare and describe changing patterns in nature that repeat themselves, such as weather conditions including temperature and precipitation, day to day and season to season.

Read an outdoor thermometer in the morning. Read the thermometer again at lunchtime and just before you leave school. Draw a thermometer each time to show the temperature. Do this for 3 days. Is there a pattern that you see?

Investigate How the Sun Warms Water

Question What can you observe about the temperature of water placed in sunlight?

Science Process Vocabulary

compare verb

When you **compare** two things, you tell how they are alike and different.

The mugs are both plastic, but they are different colors.

observe verb

When you **observe** something, you use your senses to learn about it.

I can feel that the cup is warm.

Materials

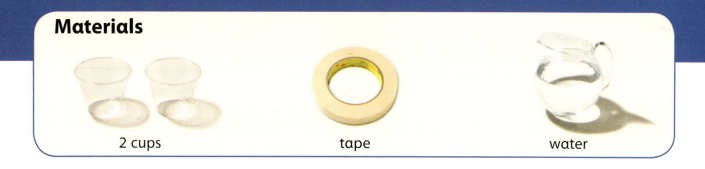

2 cups   tape   water

What to Do

1 Label 1 cup **Sunlight.** Label the other cup **No sunlight.**

2 Put the same amount of water in each cup.

Next Generation Sunshine State Standards
SC.2.N.1.4 Explain how particular scientific investigations should yield similar conclusions when repeated.
SC.2.E.7.2 Investigate by observing and measuring, that the Sun's energy directly and indirectly warms the water, land, and air.

3 Hold a cup in each hand. **Compare** the temperatures of the 2 cups. Write your **observations** in your science notebook.

4 Place the **Sunlight** cup in sunlight. Place the **No sunlight** cup in the shade.

5 Wait 30 minutes. Then hold a cup in each hand. **Compare** their temperatures. Write in your science notebook.

Record

Write in your science notebook.
Use a table like this one.

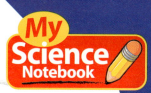

	How the Sunlight Cup Feels Compared to the No Sunlight Cup
Step 3	
Step 5	

Share Results

1. Tell what you did.

> I compared the _____ of _____.

2. What happened to the temperature of the water in sunlight?

> The water in the Sunlight cup became _____ than the water in the No sunlight cup.

Directed Inquiry

Next Generation Sunshine State Standards
SC.2.N.1.3 Ask "how do you know?" in appropriate situations and attempt reasonable answers when asked the same question by others.
SC.2.E.7.3 Investigate, observe, and describe how water left in an open container disappears (evaporates), but water in a closed container does not disappear (evaporate).

Investigate Water

Question What happens over time to water in an open cup and in a covered cup?

Science Process Vocabulary

predict verb

When you **predict,** you use what you have observed or learned to say what will happen.

I predict that the water will get warmer.

compare verb

When you **compare** two things, you tell how they are alike and different.

The two containers are different colors.

Materials

marker   2 cups with water   safety goggles   plastic wrap   rubber band

What to Do

1 One cup with water will be covered. The other cup with water will be open. **Predict** what will happen to the water in each cup.

2 Draw a line on each cup to show how much water there is.

3   Put on your safety goggles. Put plastic wrap over 1 cup. Place a rubber band around the plastic wrap. Do not put anything over the other cup.

4   Put both cups in a sunny place.

5   The next day, **observe** how much water is in each cup. Is the water in each cup still at the line you marked in step 2? Also **observe** the plastic wrap over the cup. Record your observations in your science notebook.

Record

Write or draw in your science notebook. Use a table like this one.

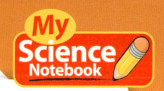

	Observations of Water
Cup with plastic wrap	
Cup without plastic wrap	

Explain and Conclude

1. **Compare** the amount of water in both cups after 1 day.

2. What do you think caused the difference?

Think of Another Question

What else would you like to find out about what happens to water left in an open container? What could you do to answer this new question?

Next Generation Sunshine State Standards
SC.2.N.1.3 Ask "how do you know?" in appropriate situations and attempt reasonable answers when asked the same question by others.
SC.2.E.7.1 Compare and describe changing patterns in nature that repeat themselves, such as weather conditions including temperature and precipitation, day to day and season to season.

Investigate How Weather Changes

Question How does the weather change from day to day?

Science Process Vocabulary

data noun

When you collect and record **data,** you write or draw what you observe.

Day 1	Day 2
sunny and warm	cloudy and cool

I collected data about the weather for two days. I recorded my data in a table.

pattern noun

When your data show that something happens the same way over and over, you can find a **pattern** in the data.

Materials

crayons

What to Do

1 What is the weather like today where you live? Is it cold, cool, warm, or hot? Record your **observations** in your science notebook.

Temperature	Cloud Cover
cool	

2 Record whether the weather today is sunny, partly sunny, cloudy, or foggy.

3 Is today's weather wet or dry? If it is wet, is there rain, freezing rain, snow, sleet, or hail? Record your **observations.**

4 Is the air today calm, breezy, or windy? Record your **observations.**

5 Collect and record the weather **data** every day for 5 days.

Record

Write or draw in your science notebook.
Use a table like this one.

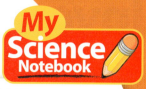

	Temperature	Cloud Cover	Wet or Dry	Wind
Day 1				
Day 2				

Explain and Conclude

1. How can you use the **data** you collected to **compare** weather for the week?

2. How did the weather change from day to day during the week? What **patterns** do you see?

Think of Another Question

What else would you like to know about how the weather changes from day to day? What could you do to find an answer to this new question?

Investigate Rainfall and Wind

Question What can you observe using a rain gauge and a windsock?

Science Process Vocabulary

observe verb

You can **observe** a windsock to see which way the wind is blowing and how strong it is.

infer verb

When you **infer,** you use what you already know to explain something.

Materials

metric ruler

tape

jar

tube with holes

2 short strings and 1 long string

paper ribbon

What to Do

1 Make a rain gauge. Use the ruler to mark centimeter lines on the tape. Start at the bottom of the tape. Number the lines from 1 to 10.

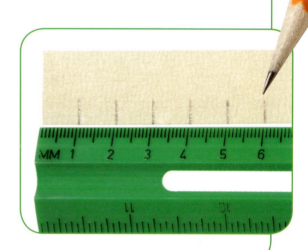

2 Put the tape on the jar. You have made a rain gauge! Choose a place outside to put the rain gauge. The next time it rains, measure how much rain falls.

The 1 cm line should be 1 cm from the bottom of the jar.

Next Generation Sunshine State Standards
SC.2.N.1.2 Compare the observations made by different groups using the same tools.

3 Now make a windsock. Tie the 2 short strings onto the tube. Make sure they make an *X*.

4 Tie the long string to the 2 short strings. Tape the paper ribbons to the bottom of the tube. You have made a windsock!

5 Choose 3 different places to test your windsock. **Observe** which way the wind is blowing and how strong it is. Record your observations in your science notebook.

No wind	
A little windy	
Very windy	

Record

Write or draw in your science notebook. Use tables like this these.

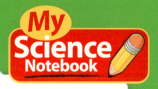

Where I Put the Rain Gauge	How Much Rain Fell

Where I Hung the Windsock	What the Windsock Looked Like	What Direction the Wind Was Coming From	How Strong the Wind Was

Explain and Conclude

1. What did you **observe** with the rain gauge? What did you observe with the windsock?

2. What can you **infer** about the weather if a windsock is not moving?

3. **Compare** the data from all groups. Do you see any **patterns** for the places where a windsock is and how strong the wind is?

Think of Another Question

What else would you like to know about observing with rain gauges and windsocks?

Next Generation Sunshine State Standards SC.2.N.1.1 Raise questions about the natural world, investigate them in teams through free exploration and systematic observations, and generate appropriate explanations based on those explorations.

Math in Science

Measuring Temperature

When scientists do some experiments, they must observe and record the temperature of something. Sometimes, they can estimate the temperature. Other times, scientists must use a thermometer to find the actual temperature.

▼ The thermometer shows the temperature of the liquid.

When scientists give the temperature of something, they tell 2 parts. There is always a number. They also tell the unit.

Thermometers measure temperature in units called degrees. Scientists use a Celsius degree. It is usually written **°C.** Some thermometers also show temperature in Fahrenheit degrees. That is written **°F.**

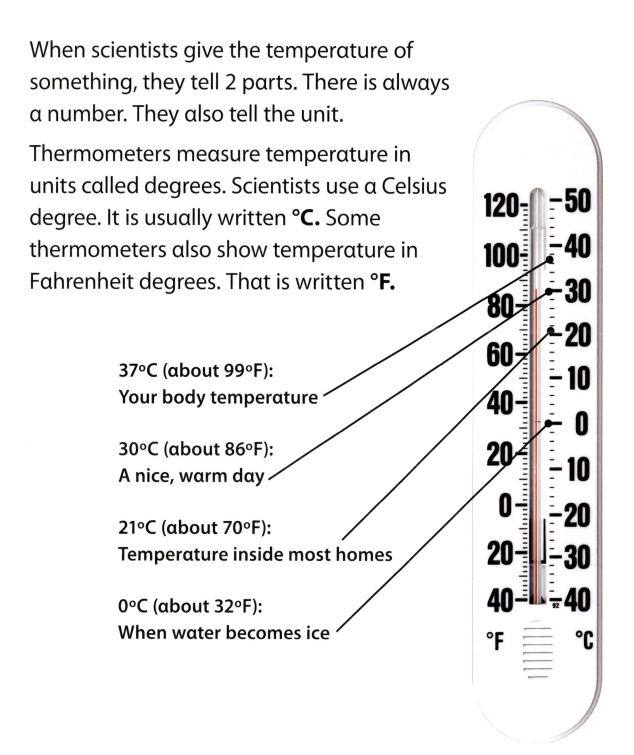

37°C (about 99°F):
Your body temperature

30°C (about 86°F):
A nice, warm day

21°C (about 70°F):
Temperature inside most homes

0°C (about 32°F):
When water becomes ice

It is easy to read the temperature on a thermometer. Look at the red liquid in the center of the thermometer. It moves down when the temperature gets colder. It moves up when the temperature gets warmer.

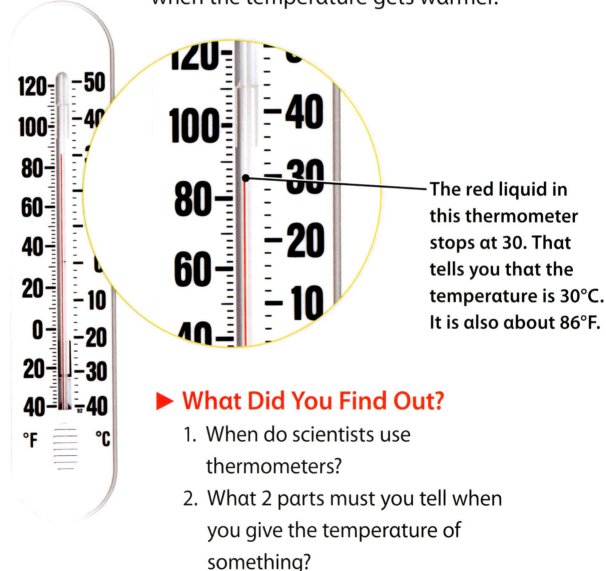

The red liquid in this thermometer stops at 30. That tells you that the temperature is 30°C. It is also about 86°F.

▶ **What Did You Find Out?**

1. When do scientists use thermometers?
2. What 2 parts must you tell when you give the temperature of something?
3. What can you do to make the red liquid move up or move down?

Measure Temperature

1. Make a table.

   - Estimate the temperature of the water for each use in the table.

   - Use a thermometer to measure the temperature of water for each use.

Use	Estimate (°C)	Measurement (°C)
Drinking		
Washing hands		
Brushing teeth		
Taking a bath		

2. Compare your estimates and your measurements.

3. Share your table with a partner and ask questions about it.

Investigate Weather Conditions

Question How can you use a thermometer, rain gauge, and windsock to observe weather?

Science Process Vocabulary

plan noun

When you make a **plan** to observe weather, you must decide what tools you will use.

measure verb

You can **measure** the temperature of the air with a thermometer.

Materials

thermometer   rain gauge   windsock

What to Do

1 With a group, make a **plan** to record the weather for 5 days. Use a thermometer, your rain gauge, and your windsock.

2 Decide where to put the thermometer. Make a list of some places you could use. Then choose 1 place.

near the flagpole
near the swings
near the basketball hoop

Next Generation Sunshine State Standards
SC.2.N.1.2 Compare the observations made by different groups using the same tools.
SC.2.E.7.1 Compare and describe changing patterns in nature that repeat themselves, such as weather conditions including temperature and precipitation, day to day and season to season.
SC.2.E.7.2 Investigate by observing and measuring, that the Sun's energy directly and indirectly warms the water, land, and air.

3 Repeat step 2 for the rain gauge and the windsock.

4 Decide when you will check each tool. You should check them the same time each day.

5 Set up your weather tools in the places you choose.

6 **Measure** the temperature, how much rain fell, and the direction of the wind and how strong it is every day for 5 days. Record your **observations** in your science notebook. **Compare** your weather **data** with those of others.

Record

Write in your science notebook.
Use a table like this one.

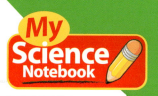

Day	Temperature	Amount of Rain	Wind Direction	Wind Strength
1				
2				

Explain and Conclude

1. If your **data** are not like those of others, give reasons why they might be different.

2. How do a thermometer, a rain gauge, and a windsock help you make weather **observations?**

▼ An anemometer like this one measures wind speed.

Think of Another Question

What else would you like to know about using tools to help you **observe** weather? What could you do to answer this new question?

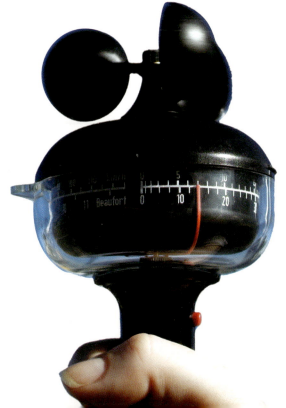

Do Your Own Investigation

Question Choose a question, or make up one of your own to do your investigation.

- What happens over time to water in open cups in the sun and in the shade?
- How does weather change from morning to afternoon?
- How does temperature in the morning compare with temperature in the afternoon?

Science Process Vocabulary

data noun

Data are observations and information that you collect during an investigation.

Open Inquiry Checklist

Here is a checklist you can use when you investigate.

Next Generation Sunshine State Standards
SC.2.N.1.1 Raise questions about the natural world, investigate them in teams through free exploration and systematic observations, and generate appropriate explanations based on those explorations.

- ☐ Choose a **question** or make up one of your own.

- ☐ Gather the materials you will use.

- ☐ Tell what you **predict.**

- ☐ Make a **plan** for your investigation.

- ☐ Carry out your **plan.**

- ☐ Collect and record **data.** Look for **patterns** in your data.

- ☐ Explain and **share** your results.

- ☐ Tell what you **conclude.**

- ☐ Think of another question.

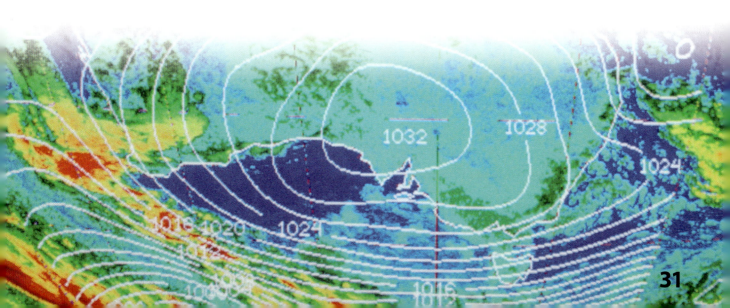

Next Generation Sunshine State Standards **SC.2.N.1.6** Explain how scientists alone or in groups are always investigating new ways to solve problems.

Science and Technology

Predicting the Weather

People have been predicting weather for a long time. Long ago some people thought a red sunset meant the weather would be nice the next day.

People also watched plants and animals for hints about the weather. They saw that some pinecones closed before it rained.

▼ **Long ago people observed that birds fly close to the ground before rain falls.**

Today some scientists predict the weather. These scientists are called meteorologists. They still look at the sky to learn about the weather. Meteorologists study the air above Earth.

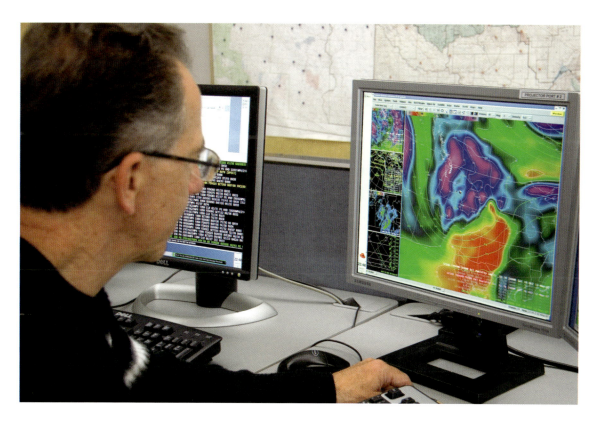

▲ **This meteorologist uses a computer to predict the weather.**

Meteorologists use weather balloons to learn about weather. Tools on weather balloons collect data about wind, temperature, and water in the air. Meteorologists use the data to make predictions.

Meteorologists also use computers to find patterns in weather data. Meteorologists use these patterns to make predictions.

► **What Did You Find Out?**

1. How did people long ago predict weather?

2. What is a weather balloon used for?

3. How do computers help meteorologists predict weather?

Predicting the Weather

1. You can use observations of clouds to predict weather.

   - Look at the pictures. Read the kind of weather each cloud brings.

   - Observe and draw the weather for 3 days. Show your school and clouds in your drawings.

   - Predict the weather for the next day. Draw your prediction.

2. The next day, compare your prediction with the actual weather.

3. Share your weather prediction drawing with others and ask questions about it.

Thunderstorms

Fair weather

Rain or snow

Featured Photos

Cover: severe weather front at twilight

Title page: lightning strikes

page 9: sunlight on ocean surface

page 13: evaporating pool in the salt flats of California

page 15: good conditions for kite flying: fair weather and moderate winds

page 31: weather map produced from satellite data, showing isobars or areas of atmospheric pressure

inside back cover: rainbow created from sunlight shining on droplets of water

ACKNOWLEDGMENTS
Grateful acknowledgment is given to the authors, artists, photographers, museums, publishers, and agents for permission to reprint copyrighted material. Every effort has been made to secure the appropriate permission. If any omissions have been made or if corrections are required, please contact the Publisher.

PHOTOGRAPHIC CREDITS
set-up photography: Andrew Northrup; stock photography: Cover (bg) Digital Vision/Getty Images; Title (bg) Digital Vision/Getty Images; 5 Michael Ventura/ Alamy Images; 6 (t) Image Source/Corbis, (c) Bubbles Photolibrary/Alamy Images, (b) terekhov igor/Shutterstock; 9 Glowimages/Corbis; 10 (t) David Arky/ Corbis, (c) image100/Corbis, (bl, br) Image Source/Corbis; 13 Image Source/ Corbis; 14 (t) Visuals Unlimited/Corbis; 15 (b) Juice Images/Corbis; 17 Purdue9394/ iStockphoto; 18 (t) Pixland/Corbis, (b) Vincent Mo/zefa/Corbis; 21 Solus-Veer/ Corbis; 22 Simon Belcher/Alamy Images; 23 Judith Collins/Alamy Images; 25 Art Vandalay/Digital Vision/Alamy Images; 26 (t) Grafton Marshall Smith/Corbis, (c) Icon Digital Featurepix/Alamy Images, (b) Ian Shaw/Alamy Images; 27 David Lewis/iStockphoto; 29 Bruce Miller/Alamy Images; 30 (t) StockTrek/Photodisc/ Alamy Images, (b) Juice Images Limited/Alamy Images; 31 Bill Bachman/Alamy Images; 32 (t) Fancy/Veer/Corbis, (b) Michael D. Kock/Gallo Images/Alamy Images; 33, 34 David R. Frazier Photolibrary, Inc./Alamy Images 35 (t) Solus-Veer/ Corbis, (c) Frank Lukasseck/Corbis, (b) Frank Lukasseck/Corbis; Inside Back Cover Creatas/Jupiterimages.

ILLUSTRATOR CREDITS Amy Loeffler

Neither the Publisher nor the authors shall be liable for any damage that may be caused or sustained or result from conducting any of the activities in this publication without specifically following instructions, undertaking the activities without proper supervision, or failing to comply with the cautions contained herein.

PROGRAM AUTHORS
Judith Sweeney Lederman, Ph.D., Director of Teacher Education and Associate Professor in the Department of Mathematics and Science Education, Illinois Institute of Technology, Chicago, Illinois; Randy Bell, Ph.D., Associate Professor of Science Education, University of Virginia, Charlottesville, Virginia; Malcolm B. Butler, Ph.D., Associate Professor of Science Education, University of South Florida, St. Petersburg, Florida; Kathy Cabe Trundle, Ph.D., Associate Professor of Early Childhood Science Education, The Ohio State University, Columbus, Ohio; Nell K. Duke, Ed.D., Co-Director of the Literacy Achievement Research Center and Professor of Teacher Education and Educational Psychology, Michigan State University, East Lansing, Michigan; David W. Moore, Ph.D., Professor of Education, College of Teacher Education and Leadership, Arizona State University, Tempe, Arizona

THE NATIONAL GEOGRAPHIC SOCIETY
John M. Fahey, Jr., President & Chief Executive Officer
Gilbert M. Grosvenor, Chairman of the Board

National Geographic School Publishing
Hampton-Brown
www.NGSP.com

Printed in Mexico.
RR Donnelley, Reynosa

ISBN: 978-0-7362-7639-9

11 12 13 14 15 16 17

10 9 8 7 6 5 4